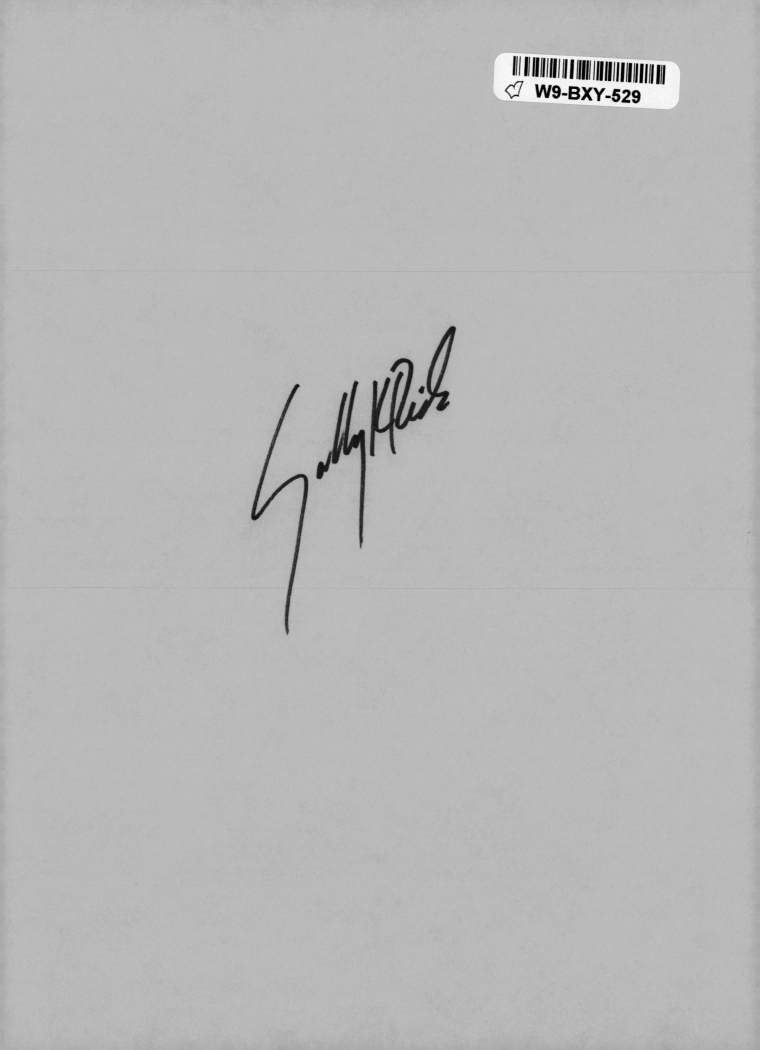

Sally Ride & Tam O'Shaughnessy

THE MYSTERY OF
MARS

Sally Ride
Science

To our sisters—Bryn, Bear & Kim—
For their friendship, humor,
and love over the years

Published by Sally Ride Science
9191 Towne Centre Dr, Suite L101, San Diego, CA 92122

Imagine a planet so far away that we see it as just a point of light in the night sky, yet so near that we dream of visiting it someday.

Gazing at Earth through the windows of the space shuttle, I looked down at the sparkling blue oceans, snowcapped mountains, and sprawling cities of the planet below. Above the horizon, I saw a familiar red point of light and wondered what it would be like to be in orbit around Mars. There, Earth would be just a pale blue speck of light, millions of miles away. The planet below me would be a rugged red world, with enormous volcanoes piercing through red clouds, and sheer cliffs dropping off into huge canyons.

→ *Earth and its moon from a distance of about 7 million miles. Our planet would look even smaller and fainter from Mars.*

↓ *Florida, through the window of the space shuttle.*

It will be many years before astronauts visit Mars. But robot spacecraft have begun to explore that distant world. Through their eyes, we see a planet that is starkly beautiful, but cold, dry, and desolate. There is no water on its surface, no oxygen in its atmosphere, and no life in its soil. But we also see clues that ancient Mars was very different. Long, long ago, Mars was a warmer planet, perhaps with thick clouds, ice-covered lakes, and maybe even primitive microscopic life.

Over the years, we have learned a lot about this mysterious planet. But we still have many questions. If early Mars was different from the planet we see today, why did it change? Did primitive life evolve on Mars? If so, have any Martian microbes survived? This book describes what we know about Mars and what we hope to learn about this intriguing planet in the future.

fascinated with the night sky as we are. When their scientists charted the heavens, they found that a few of the twinkling lights seemed to wander across the sky against the background of stars. These were called planets, which is from the Greek word for "wanderers." One of the brightest had a reddish color and was later named Mars, after the Roman god of war.

Many hundreds of years later, scientists painstakingly plotted the path of Mars across the sky and learned that this planet, like Earth, orbits the sun. Their observations showed that Mars travels around the sun in about two years and that Mars is Earth's neighbor in the solar system. Earth is the third planet from the sun. Mars is the fourth.

With the invention of the telescope in the 1600s, astronomers had a much better view of Mars. Instead of seeing just a twinkling reddish light, scientist got a blurry view of a large, round world—maybe a world like Earth.

They measured the size of the image in their telescopes and discovered that Mars was a smaller planet than Earth. The distance across Mars at the equator is about half the distance across Earth.

As telescopes improved, the blurry disk came a bit more into focus. The north and south poles of Mars, like those of Earth, appeared to be covered with icecaps. And some parts of the Martian surface were definitely darker than others. As astronomers watched these markings through their telescopes, they noticed that this planet, like Earth, was rotating—spinning like a top. They timed the rotation and found that Mars spins around once every 24 hours and 37 minutes, which means that a day on Mars is just a little longer than a day on Earth.

The size and shape of the fuzzy, dark areas change with the Martian seasons. Some astronomers thought that these areas looked vaguely green. Could they be vegetation? If so, could there be other forms of life—maybe even civilizations—on this faraway world?

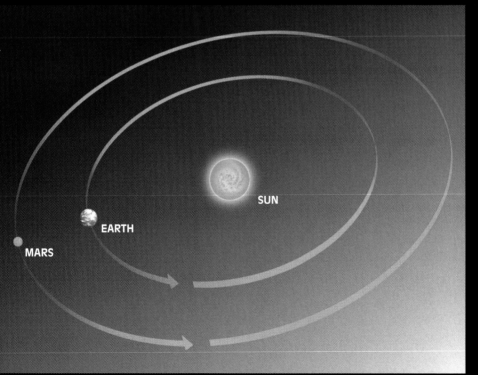

← Mars' orbit is approximately one and a half times the size of Earth's. (For clarity, the inner planets, Mercury and Venus, are not shown. The sizes of the Sun and planets are not to scale.)

SUN

EARTH

MARS

n Mars, the sun rises in the morning, climbs in the ky, and sets in the evening, just as it does on Earth. This is because Mars, like Earth, rotates about its xis while it's traveling around the sun. To a person tanding on the rotating planet, it looks as if the sun is noving across the sky.

A day on Earth is 24 hours long because it takes 24 ours to make one complete rotation. Mars rotates at bout the same rate as Earth. It takes Mars 24 hours nd 37 minutes to spin around once. A day on Mars, alled a sol (the Latin word for "sun"), is 37 minutes onger than a day on Earth.

On some other planets, a "day" is very different. A day on Venus is very long—244 Earth days! This is because Venus rotates very slowly, only once every 244 days. A day on Jupiter is very short—only 10 hours ong—because Jupiter rotates very quickly.

AS MARS SPINS, A PARTICULAR PLACE ON ITS SURFACE MOVES FROM DARKNESS TO DAWN, AND THEN TO DAYLIGHT.

A Year On Mars

A year on Earth—365 days—is the time it takes our planet to travel around the sun. Mars is farther from the sun, so it takes longer to go around. A year on Mars is 687 Earth days, nearly two Earth years.

On Earth, each year has four seasons: spring, ummer, fall, and winter. We have seasons because Earth's rotation axis is tilted. When Earth is at (A) in it's orbit around the sun, the sunlight shines more directly on the northern half of the planet. A city north of the equator is in daylight for more than 12 hours a day. It s summer in the Northern Hemisphere and winter in the Southern Hemisphere.

Six months later, the Earth is halfway around in its orbit (B). The sun shines more directly on the southern half of the planet. At this time of year, it is summer in the Southern Hemisphere and winter in the Northern Hemisphere.

Earth's axis is tilted about 23.5 degrees. Mars s also tilted, about 25 degrees, nearly the same as Earth. So Mars also has a winter, spring, summer, and fall. The biggest difference is that since a Martian year s twice as long as an Earth year, each Martian season

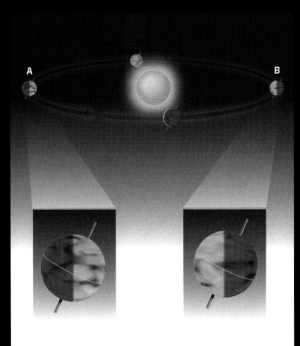

Giovanni Schiaparelli turned his telescope toward Mars. He described features on the surface as *canali*, the Italian word for "channels," but this description was incorrectly translated into English as "canals." Soon astronomers from around the world, looking at a blurry Mars, were sketching maps showing long, straight "canals" criss-crossing the planet's surface. But their imaginations were creating patterns in the blurry images that were not really there.

The most famous of these astronomers was an American named Percival Lowell.

believed he saw on the surface of Mars. He also developed a theory that the long canals were built by intelligent Martians. H thought they were trying to save their pla from a terrible drought by pumping water from the polar regions to the dry farmland near the equator.

Few scientists shared this view. But Low wrote about his ideas in magazines and newspapers. By the early 1900s, much of the public had heard about Lowell's race of intelligent Martians and their system of water canals.

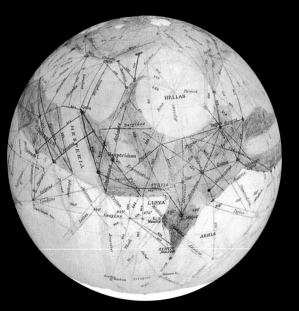

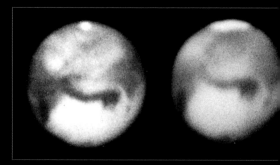

↑ Mars looked blurry through early telescopes. But astronomer could see the light and dark areas change with the seasons.

← Percival Lowell sketched serveral globes of Mars. This one, dr in 1907, shows the canals he believed he saw.

↓ This map was made in the early 1960s. It was the most accu map of Mars before spacecraft surveyed the planet.

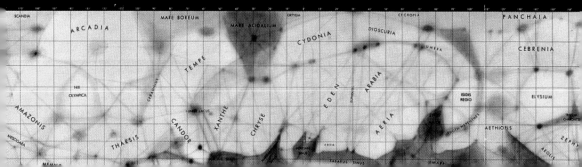

Over the years, telescopes became better and better. But the telescopes were viewing a planet millions of miles away and could not see the Martian surface clearly. By the early 1960s, the most accurate globe of Mars still showed very little detail. It even included a few straight lines that looked like Lowell's canals.

But space exploration was just beginning. In 1965, the *Mariner 4* spacecraft flew past Mars and sent back the first close-up photographs of its surface. The photos showed no vegetation, no canals, and no water. Overnight, our view of Mars changed.

↑ *The* Mariner 4 *spacecraft.*

↓ *One of the 22 photographs taken by the* Mariner 4 *spacecraft as it flew past Mars. This was the first photograph clearly showing craters on the Martian surface.*

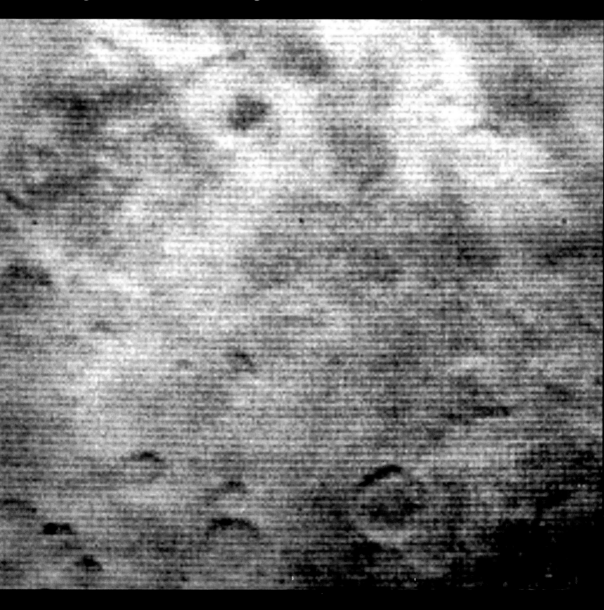

No person has ever set foot on the Red Planet. Most of what we know about Mars we have learned by sending robot spacecraft on one-way trips to that faraway world.

The first spacecraft flew past Mars, transmitting information to Earth as they zipped by. Later spacecraft went into orbit around Mars. As they circled the planet, they sent pictures and data millions of miles back to Earth. Only a few spacecraft have actually landed on Mars. Acting as our remote senses, they have sniffed the Martian air, gazed at Martian rocks, and sifted through Martian soil.

The spacecraft circling Mars have sent back thousands and thousands of photographs. From these photographs, scientists have pieced together a detailed mosaic of the planet's surface.

Even at a glance, it is obvious that Mars is very different from Earth. When you compare pictures of the two planets, the first thing you notice is that Earth is covered with sparkling blue water. Mars has none. There are no Martian oceans. There are no rivers or lakes. There is no liquid water on the surface at all.

Earth and Mars, to scale.

Artist's drawing of the Mars Global Surveyor *in orbit around Mars.*

Mars Global Surveyor

Back in the days of Schiaparelli and Lowell, maps of Mars were drawn by hand. Astronomers squinted through telescopes and sketched the fuzzy features that they saw. Now, a century later, the mountains, craters, and canyons on Mars are being charted in accurate detail by orbiting spacecraft. Mars Global Surveyor began mapping the planet in 1999. As it circles above, its high-resolution camera snaps photographs of the surface below.

Meanwhile, the spacecraft's laser altimeter charts the highs and lows of the planet's surface. This instrument measures elevation by beaming laser pulses down to the ground, then timing how long it takes them to bounce back up to the spacecraft. Scientists combined the images and surface heights to make the first accurate global map of Mars. The map provides valuable scientific information and is used to select landing sites for future spacecraft.

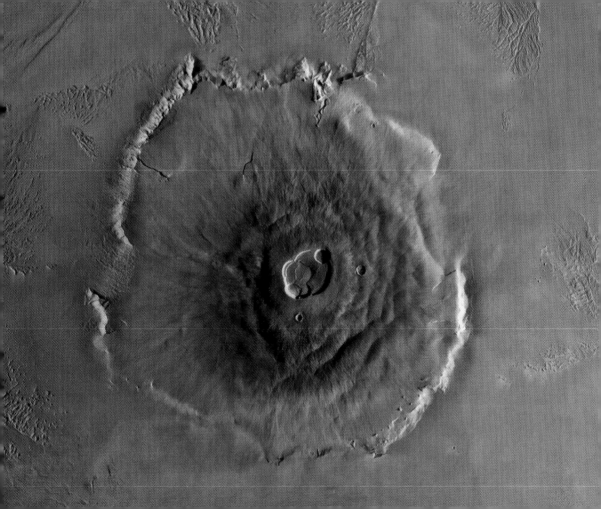

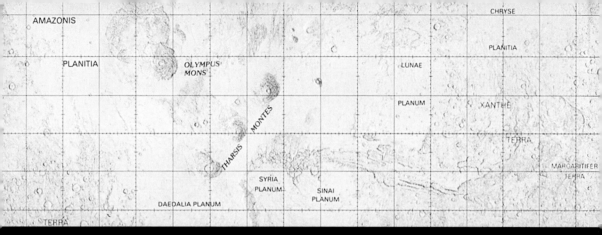

↑ *Map showing Olympus Mons and a line of three other huge volcanoes in the Tharsis region (the Tharsis Montes: Arsia Mons, Pavonis Mons, and Ascreus Mons). The map also shows Valles Marineris and Chryse Planitia, the region where both* Pathfinder *and* VIking 1 *landed.*

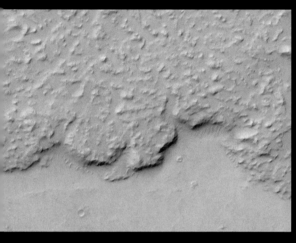

↑ *Olympus Mons is much larger than Mauna Loa in Hawaii, the largest volcano on Earth.*

↖ *The edge of a huge lava flow southwest of Arsia Mons, a large volcano in the Tharsis Region.* (Mars Global Surveyor)

↓ *Olympus Mons rises above the Martian clouds.* (Viking)

The bulging Tharsis region has split and cracked in many places, creating networks of deep fractures. Valles Marineris (Mariner Valley, named after the *Mariner* spacecraft that first photographed it) is like a big gash in the planet. This huge valley is the longest and deepest in the solar system. Valles Marineris would stretch all the way across the United States, from the Pacific Ocean in the west to the Atlantic Ocean in the east. The entire Grand Canyon would fit in one of its small side canyons. If a Martian rock tumbled off the rim of Valles Marineris, it would fall four miles before hitting the canyon floor.

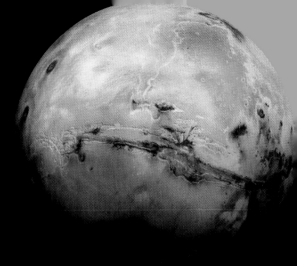

↗ *Valles Marineris stretches across the middle of the planet. Two of the Tharsis volcanoes, Pavonis Mons and Ascraeus Mons, are visible on the far left.*

↓ *A computer-generated image, looking down Valles Marneris.*

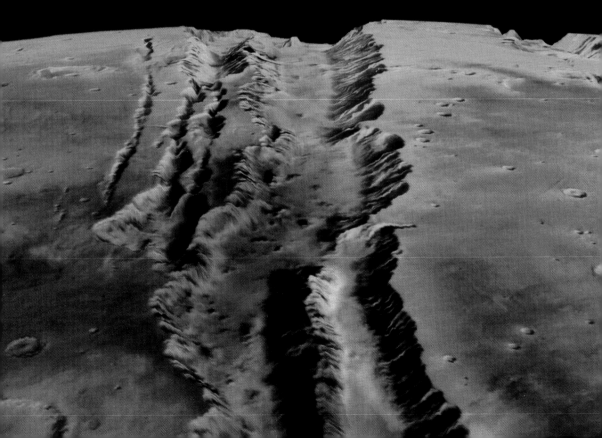

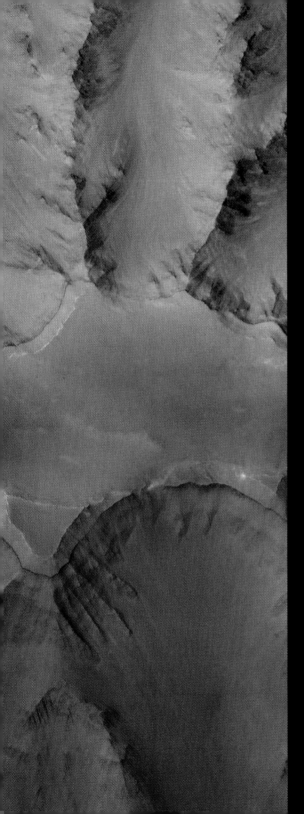
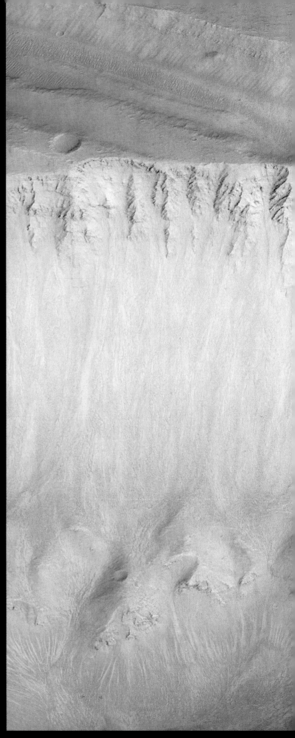

High-resolution views of different parts of Valles Marineris. Left: Six-mile section of a plateau (center of the image) that forms a step in the cliff. Right: Part of a steep canyon wall. Layers of rock are visible in the wall; landslides have helped shape the canyons. (Mars Global Surveyor)

Photographs of Mars show that much of the planet is covered with craters. Early in the history of the solar system, Mars, Earth, and the other planets were bombarded with meteors that blasted craters in their surfaces. Most of Earth's craters have been erased. Some have been eroded away by water, some buried as the land has changed, and some hidden by vegetation. But Mars still shows the scars of thousands of these violent collisions. The largest is the Hellas Basin, a huge depression that is 1,300 miles across and six miles deep. The crater, which resulted from a truly planet-jarring collision long ago, would cover nearly half the United States. Four billion years ago, all of Mars was probably covered with craters. Since then, much of the land north of Mars' equator has been flooded by lava flowing from the volcanoes in the Tharsis region and elsewhere. Many of the old impact craters in the north have been covered over, buried by lava.

In contrast, the southern part of Mars is still carpeted in craters. This part of the planet has not been resurfaced by flowing lava. The heavily cratered land is nearly four billion years old and has been changed very little during that time.

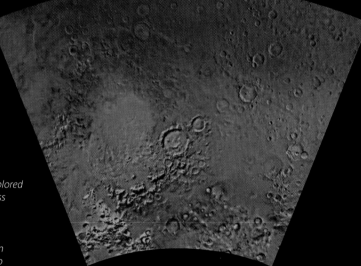

→ *The heavily cratered Martian highlands. The circular light-colored area is the Argyre impact basin, a crater nearly 500 miles across that is surrounded by a rim of rugged mountains.* (Viking)

↓ *Not many of Earth's impact craters are still visible. The Manicouagan Crater in Quebec, Canada, was formed 200 million years ago and is now heavily eroded. Though small compared to Argyre, it is almost 45 miles across.*

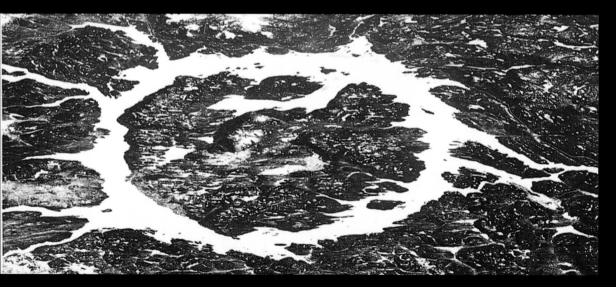

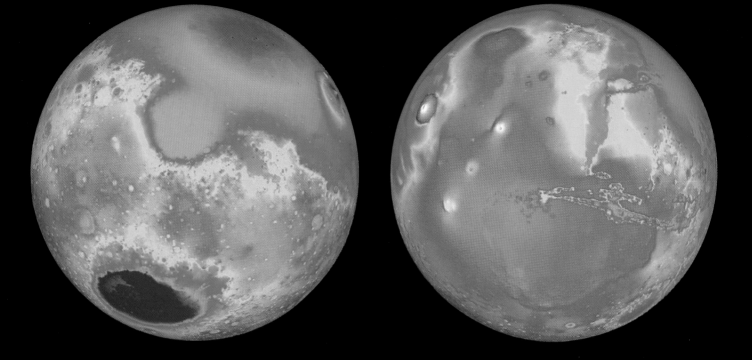

↑ These maps of Mars were made from millions of elevation meansurements taken by Mars Global Surveyor. The colors represent the different heights of the surface. The lowest parts of the planet are purple and blue; the highest parts are white and red. Mars has the highest, the lowest, and the smoothest land in the solar system. Above Left: The huge Hellas Basin (the large dark blue circle), surrounded by countless smaller craters. Above Right: The Tharsis region with its large volcanoes (white) and Valles Marineris (horizontal green gash). The largest volcano is Olympus Mons.

↓ The flat map shows the dramatic difference between the northern and southern parts of the planet. The south is high and scarred with craters; the north is about three miles lower and very smooth. Why is the north so smooth? Part of it has been paved over by lava, and part may once have been covered by oceans.

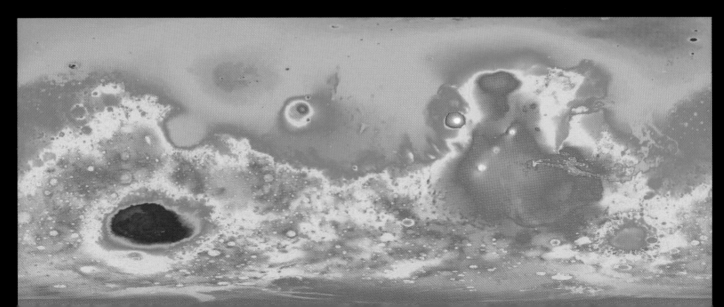

In 1976 *Viking 1* and *Viking 2* settled softly onto the surface of Mars. They were the first spacecraft from Earth ever to visit the Red Planet. Twenty-one years later, *Pathfinder* dropped out of the Martian sky to join them. A parachute opened to slow it down, then giant air bags inflated to cushion it during impact. *Pathfinder* bounced hard more than 15 times before it rolled to a stop on the red Martian soil.

Although the *Viking* and *Pathfinder* landers arrived at different locations, they landed in similar terrain. Engineers did not want to risk landing these precious spacecraft on the edge of a cliff or the side of a volcano. They guided them to different sites on the gently rolling Martian plains. The pictures the spacecraft sent back showed flat, windswept landscapes strewn with gray rocks and covered with fine red dust.

↑ *The* Pathfinder *lander. When the airbags that had protected it deflated,* Pathfinder *opened like a flower to reveal a camera, a weather station, and its rover,* Sojourner. *This photograph was taken by* Sojourner *after it had left the lander. The lander's camera, at the top of the mast, is looking at* Sojourner.

The two *Viking* landers could not move. Their robot arms could only reach out a few feet to scoop up small samples of soil. *Pathfinder* carried the first rover to Mars. The rover, *Sojourner*, was about the size of a small dog. It moved at a snail's pace, but was able to travel several yards from the lander.

The little robot geologist dug its wheels into the red Martian dirt, churning it up to analyze the soil. It roamed through a garden of nearby rocks, ranging in size from pebbles to boulders.

↑ *Sojourner had TV cameras for eyes and was steered by drivers back on Earth.* (Pathfinder)

↓ *Pathfinder* landed in Ares Vallis, an ancient floodplain. Many of the rocks here were deposited by floods billions of years ago. This panorama also shows Pathfinder's *deflated airbags* and the ramp that its small rover, *Sojourner*, drove down to reach the surface. The rover is analyzing a rock a few feet from the lander. When *Sojourner* rolled down Pathfinder's *ramp*, it become the first rover ever to explore the Martian surface.

Two larger, more capable rovers arrived seven years later. *Spirit* and *Opportunity* bounced onto the Red Planet three weeks and six thousand miles apart in early 2004. Their mission was to look for proof that there was once water on Mars. Their landing sites were carefully chosen. Each was a place that scientists suspected was covered with water long ago.

Opportunity landed in Meridiani Planum, where orbiting spacecraft had detected a mineral that often forms in water. *Spirit* landed in a crater that scientists thought might have once been an ancient lake.

Spirit and *Opportunity* were the size of small dune buggies, not small dogs! Each could roam a few miles, not just a few yards. Each carried cameras, sensors, and magnifying glasses. And each could grind samples of rocks and analyze the chemistry of the rocks and soil. They spent over two Earth years—one entire Martian year—driving up Martian hills and down into small craters. Each of the robot backpackers trekked over four miles.

↓ *This panorama was made from over 40 images. They were taken by the rover* Opportunity *from the bottom of the 65 foot deep Endurance crater. The layers of rock provide evidence for many cycles of evaporation and reappearance of water.*

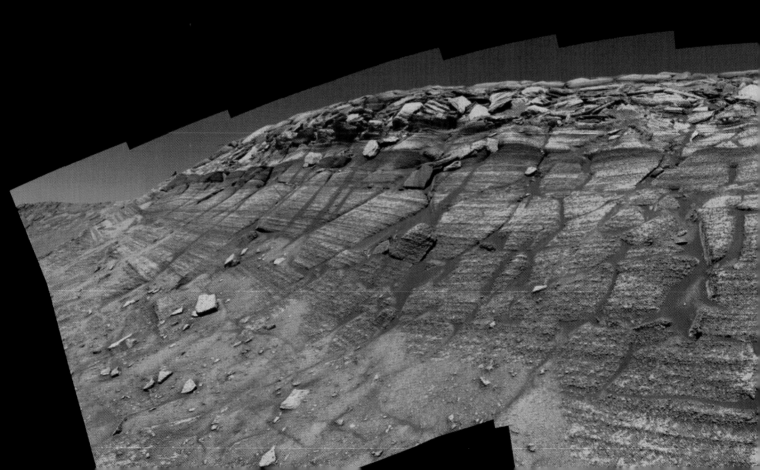

→ Artist's drawing of one of the twin Mars Exploration Rovers. Spirit *and* Opportunity *landed on different parts of the planet, and each traveled miles across the Martian surface.*

By sheer luck, *Opportunity* landed in a small crater in Meridiani. On the inside wall of the crater, the rover found exposed layers of old rock. When scientists back on Earth studied the data, they found that the rock was made of layered sediments. That told them that the area was once a salty sea. When the water evaporated, a salty sediment was left behind. The rover's photos showed many layers, so the waters must have come back, then evaporated, time and again—leaving behind a new layer each time.

Spirit traveled from its landing site to a range of hills off in the distance. On one hillside it found an exposed pile of layered rocks. Sure enough, scientists could tell that those rocks had been changed by water. The land had once been soaking in water.

The adventures of *Spirit* and *Opportunity* showed scientists that Mars did have water—lots of water—billions of years ago.

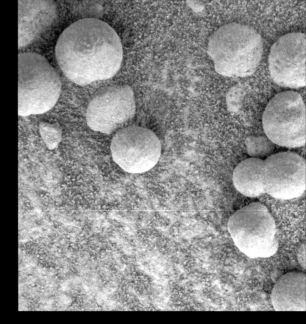

↑ *These small spheres, about the size of BB's, are made of iron oxide. They formed long ago in the presence of water. Scientists called them blueberries because they look so much like them.* (Opportunity)

↓ *As the rover* Opportunity *heads off to explore Mars, it looks back at the small crater where it landed. The tracks show its path out of the crater. This area may have once been the shore of a salty sea.*

For decades scientists have wondered whether there might be microscopic life on Mars. The *Viking* landers were the only spacecraft that carried experiments to try to answer that question. Their long robot arms scooped up samples of Martian soil and brought them inside the landers. Then instruments analyzed the red dirt for signs of life.

In planning these experiments, scientists assumed that Martian microbes would be similar to those on Earth: they would take in food molecules, grow, and release waste molecules. In one experiment, nutrients were added to soil samples. Then an instrument looked for carbon dioxide gas, which might signal that living organisms had eaten the food.

This experiment did not find evidence of Martian life. Results from the other two experiments were also negative.

But are the building blocks of life present in the Martian soil? Another experiment looked for organic molecules, the molecules that make up living things. Samples of soil were heated, and instruments watched for gasses that would be released if organic molecules were present. It was a great surprise when none were found. Scientists know that meteorites and interplanetary dust deliver a steady supply of organic molecules to the Martian surface. So even if there are no living organisms, there should still be some organic molecules. Scientists now suspect that they are being destroyed by harsh chemicals in the Martian soil.

Spacecraft that have followed *Viking* have not carried experiments to look for evidence of life. The few *Viking* experiments, at two tiny spots on Mars, are all scientists have to go by. But many believe that it is possible that primitive life exists beneath the surface, or that life existed on the planet long ago.

↓ *The* Viking 2 *lander's robot arm scoops up a sample of soil and leaves its mark in the ground.*

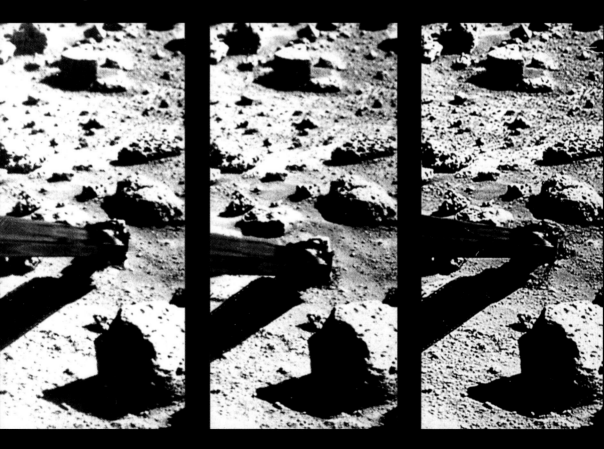

Earth is surrounded by an atmosphere that protects all the plants and animals on the planet from extreme conditions in space. It shields us from the sun's radiation, helps keep our planet warm, and contains the oxygen that many of Earth's creatures need to survive.

Mars, too, has an atmosphere, but it is very different from Earth's. The Martian atmosphere is very, very thin and is made up almost entirely of carbon dioxide. Fine red dust fills the thin air and creates a pink sky all year round.

Some of the landers set up small weather stations. The stations radioed Martian weather reports back to Earth. Like the weather on Earth, the weather on Mars changes from day to day and from season to season. On some days the pink sky is mostly sunny, with light winds and wispy rose-colored clouds. On other days the sky is overcast, with strong winds and swirling cinnamon-colored dust.

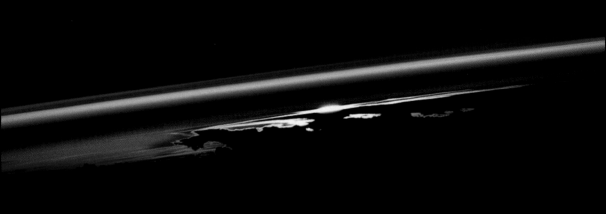

↑ Space shuttle astronauts took this picture of Earth's atmosphere at sunset. Storm clouds rise about eight miles above the planet's surface.

↓ An unusually clear view of the Martian atmosphere. Thin layers of haze extend 25 miles above the horizon. (Viking)

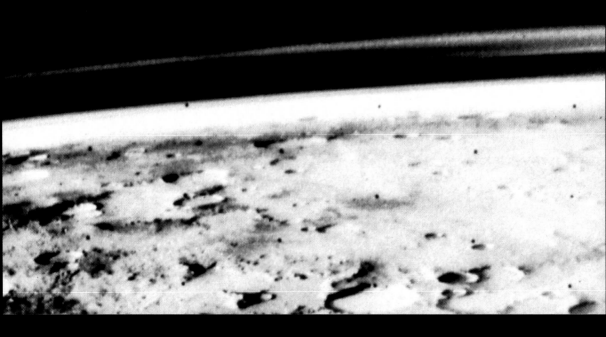

There is very little water vapor in the Martian
atmosphere. Martian clouds contain crystals
of water ice, but the air is too thin and
too cold for raindrops to form. In the early
mornings, a thin veil of fog might fill the
distant canyons, but there is no dew on the
canyon walls. The rain that nourishes all life
on Earth never falls on Mars.

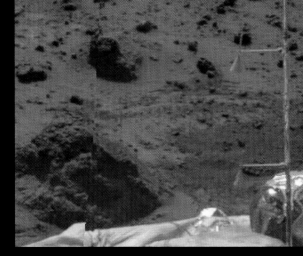

→ Pathfinder's *weather station. The windsocks on the far right are
slightly tilted because they are being blown by the Martian wind.*

↓ *During the late afternoon, clouds accumulate around and above
Olympus Mons.* (Mars Global Surveyor)

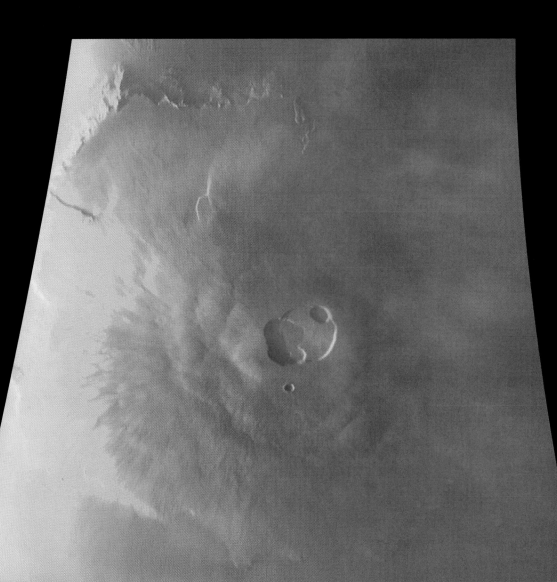

The air on Mars is very thin. Because it is so thin, water cannot exist as a liquid on the surface. If an astronaut on Mars poured a glass of water, it would soon boil away.

The boiling point of water (the temperature at which it turns into gas) depends on the pressure of the surrounding air. You can see this yourself if you go camping in the mountains. Near sea level, water has to be heated to 212 degrees Fahrenheit before it will boil. As you climb up a mountain, the air gets thinner and thinner, so water boils at a lower and lower temperature. On a 5,000 foot mountain (and in the mile-high city of Denver), water boils at about 203 degrees Fahrenheit. At the top of Mount Everest, the highest mountain on Earth, water boils at only about 160 degrees Fahrenheit.

When spacecraft measured the air pressure on the surface of Mars, they found that it is the same as it would be on a mountain more than 3 times as high as Mount Everest. When the air is that thin, water boils at very low temperatures—temperatures near its freezing point. That means that water on Mars exists either as ice or as water vapor (a gas), but not as a liquid.

Mars is very, very cold. Even on bright summer days, temperatures may only reach 10 degrees Fahrenheit—22 degrees below the freezing point of water. When the sun goes down, the temperature falls to a frigid 110 degrees below zero. Earth's atmosphere helps keep our planet warm overnight. But on Mars, the atmosphere is so thin that after the sun sets, the planet's heat quickly escapes to space.

If you were standing on Mars on a summer morning, your feet would be warm but your ears would be freezing! As the sun warms the soil, the air a few inches above the ground is heated to nearly 50 degrees Fahrenheit. But just a few feet off the ground, the temperature plummets.

Winters on Mars are so cold that nearly 20 percent of the planet's air actually freezes out of the sky. Carbon dioxide gas in the air turns to ice and is trapped in Mars' polar icecaps until spring. Then when the temperature warms, the carbon dioxide goes back into the air as a gas.

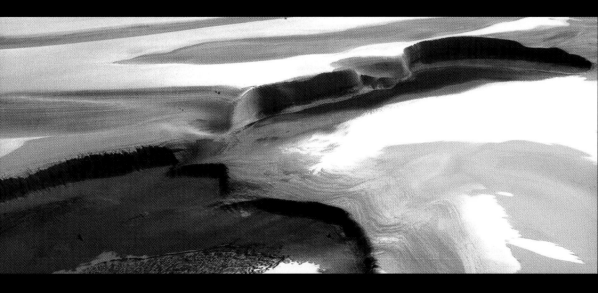

← Wispy clouds about 10 miles high, made of water ice condensed on particles of red dust. (Pathfinder)

↓ Frost covers the Martian landscape near the Viking 2 landing site in Utopia Planitia in the Elysium region.

↑ The north polar ice cap of Mars has layers of dust and water ice. The cliffs in this image are over one mile high. (Mars Express)

Mars is a windy planet. Dust devils whirl across the surface, lifting red dust high into the sky. During some parts of the year, ferocious winds stir up huge dust storms in the Southern Hemisphere that can grow to cover the entire planet. These dust storms are far worse than any on Earth and can completely block our view of the planet's surface for weeks at a time.

Over the ages, Martian winds have created sand dunes over much of the planet. Some dunes appear to be ancient remnants of an earlier time when the air was thicker and the wind could more easily blow sand around. Other dunes are still active today.

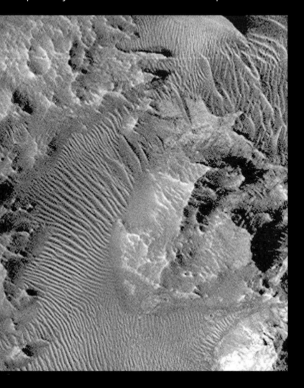

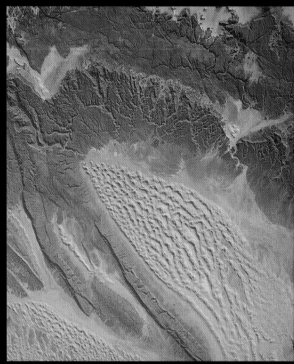

↑ Sand dunes like these cover much of Mars. (Mars Global Surveyor)

↓ The Spirit rover captured this photo of two dust devils whirling across the surface of Mars. Dust devils like these extended the rover's life when they swept over the rover and cleaned off its solar panels.

↑ Sand dunes are also common on Earth. These dunes in Algeria were photographed by astronauts in the space shuttle.

↑ *A Martian sunset.* (Spirit)

When the first astronauts visit Mars, what will they find? Though an astronaut could not survive without a spacesuit, she would feel more at home there than anywhere else in the solar system. She could stand on a rocky surface, scoop up a gloveful of dirt, and explore extinct volcanoes and ancient canyons.

She would need the spacesuit to protect her from the thin Martian air and the extreme cold. The spacesuit would be bulky, but not heavy. Because Mars is smaller than Earth, the pull of gravity on its surface is lower. She and her spacesuit would weigh about one-third

As she hiked across the rugged, rocky terrain, her boots would leave deep footprints in the dusty soil. Fine red dust would cling to her spacesuit. Even on days when the wind was calm, she would look up at a pink sky loaded with red dust. As she headed back to the warmth of her spacecraft at the end of the day, she would look past the silhouettes of crater rims at a dimmer setting sun.

The planet she was exploring would seem strangely familiar. But it would be missing the air and water that make Earth habitable, and the plants and animals that share her

Mars is a dry, desolate planet.
But the same photographs that show us the dry desert of today also let us look back in time, giving us a glimpse of an earlier Mars. The old, cratered terrain is etched with dry riverbeds. This suggests that water once flowed on the planet's surface. Today the air is far too thin, and far too cold, for streams to flow in the Martian valleys. But 3.8 billion years ago, when Mars was young, there was water. Ancient Mars was very different from the planet that now twinkles in our night sky.

Mars and Earth formed at the same time, about 4.5 billion years ago. Before then, our entire solar system was just a huge rotating cloud of gas and dust. Then the cloud began to collapse. Its center shrank to form our sun. As it contracted, the sun heated up. Soon the temperature and pressure at its center were high enough for colliding hydrogen atoms to combine to form helium. This process releases enormous amounts of energy and is the source of the sunlight that bathes our solar system. The sun began to shine.

↓ → *Viking and* Mars Global Surveyor *photographed this Martian valley cutting through a region in the cratered southern highlands.* Mars Global Surveyor's *higher-resolution view (right) was the first to show a small inner channel (at the top of the photo). This inner channel suggests that a steady flow of water may have cut the valley. This is the way that many river valleys are formed on Earth.*

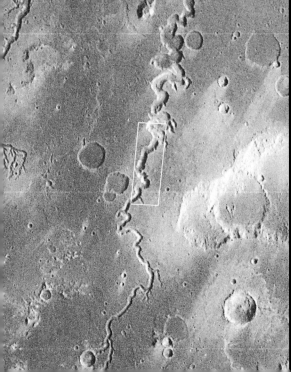

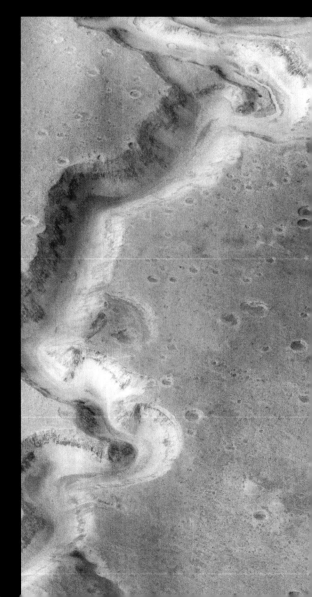

slowly, relentlessly rearranging the continents and oceans. As they collide, they drive part of the crust down into Earth's molten interior and recycle its rock.

Mars is a smaller planet, so it cooled more quickly. Its earliest crust may have broken into plates, but as the planet cooled its crust thickened. Mars soon became a one-plate planet. Still, its hot interior was active. The Tharsis bulge was pushed up by molten rock, and Mars' enormous volcanoes were all driven by heat from below.

Earth is still a very active planet. Earthquakes and volcanoes, caused by the motion of plates, occur every day. Mars began with less internal heat and lost it more quickly. As the planet's hot interior cooled, it had less and less effect on the surface. Most of the southern half of Mars hasn't changed in nearly four billion years.

Earth's ancient atmosphere also began to change. The carbon dioxide was slowly removed from the air and stored in its oceans and land. Early torrential rain carried this gas, dissolved in raindrops, down to the ground. Some splashed in the early oceans. Some rained onto the land, and slowly dissolved the rock it trickled over. The dissolved chemicals could then recombine to make new rock, called carbonate rock. This process takes carbon dioxide that was once in the atmosphere and traps it in rock.

Some of this carbon dioxide eventually returns to the atmosphere. As Earth's plates move, rock is driven down into the planet's hot interior. When the molten rock rises again through volcanoes, carbon dioxide bubbles back out into the air.

Mars' atmosphere changed too. Carbon dioxide may have been gradually pulled from the atmosphere to the land, just as it was on Earth. Some might have come down dissolved in Martian rain. As on Earth, the carbon dioxide in water would dissolve bits of rock, then combine with these minerals to form carbonate rock.

Early in Mars' history, while volcanoes were active, some carbon dioxide was sent back into the atmosphere. But eventually, as its insides cooled, this nearly stopped.

Also, because Mars is smaller than Earth, it did not have quite as tight a grip on its atmosphere. The lightest gases could escape Mars' gravitational pull more easily than they could Earth's. As sunlight struck molecules high in Mars' atmosphere, it broke some of them apart into atoms. Those atoms could then drift away into space. For example, nitrogen remained in Earth's atmosphere but was slowly, atom by atom, able to escape from Mars' atmosphere.

Mars lost most of its atmosphere to space, and some became locked in the land. If Mars had been the same size as Earth, it might still have a thick atmosphere. Instead, it now has very little of its original atmosphere left.

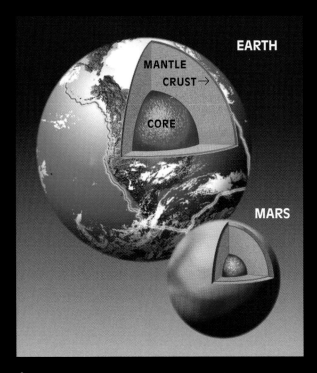

↑ Both Earth and Mars have solid cores surrounded by a molten mantle of liquid rock and a thin rocky crust. Mars is a smaller planet. It cooled more quickly and has a smaller core and a thicker crust.

← An early planet, as it might have looked about 4.4 billion years ago. This planet could be either Earth or Mars at that time.

Life first appeared on our planet shortly after Earth's crust formed. Fossils of primitive microscopic cells date back 3.5 billion years. There is even some evidence that life may have existed as early as 3.9 billion years ago.

How could life begin on such a violent, stormy, meteor-battered planet? Amazingly, life has only a few basic needs—water, organic molecules, and an energy source. Each of these was plentiful on the young Earth.

Water is essential to life. It makes up about two-thirds of every living thing, from bacteria and butterflies to earthworms and elephants. The watery environment inside cells is where all the chemical reactions take place that keep living things alive. Water covered Earth long ago, just as it does today.

The elements carbon, hydrogen, nitrogen, and oxygen are also vital. These are the most common elements in the solar system. Simple molecules, made of combinations of these atoms, were abundant in the waters of early Earth. Then some form of energy—maybe sunlight, lightning, or heat from inside the planet—caused these elements to combine into more complex organic molecules, the molecules that make up living things.

Over time, the first living organisms formed out of the brew of organic molecules in the waters of Earth. These were microscopic living things that were able to reproduce themselves from the organic molecules in their environment. But this process was not perfect. By chance, some organisms were better able to compete for the necessary organic molecules than others. Evolution had begun.

No one knows for sure when or where the first cells formed. But scientists think there were several places on early Earth where life could have begun. It may have started on the ocean floor, where organic molecules settled out of the water near volcanic vents. It may have started in volcanic hot springs, where the rising water was enriched with minerals as it gushed toward the surface.

The earliest forms of recognizable life were tiny single cells that resemble today's bacteria. And for most of Earth's history— until about 500 million years ago—our planet was inhabited only by these microscopic organisms, which are much too small to see. For more than 3 billion years, life on Earth was invisible. Earth's surface must have looked as barren as the Martian landscape does today.

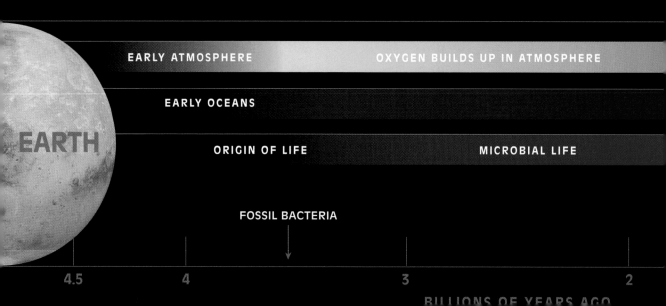

EARTH

EARLY ATMOSPHERE   OXYGEN BUILDS UP IN ATMOSPHERE

EARLY OCEANS

ORIGIN OF LIFE   MICROBIAL LIFE

FOSSIL BACTERIA

4.5   4   3   2

BILLIONS OF YEARS AGO

↑ Painting of Earth as it might have looked about 3.8 to 4 billion years ago.

← Earth time line from 4.5 billion years ago to the present, showing the approxmiate times of some important periods in Earth's evolution.

OZONE LAYER FORMS

MULTICELLULAR LIFE

HUMAN BEINGS

LAND ANIMALS

LAND PLANTS

1   0.5   now

About four billion years ago, ancient Mars may have been similar to ancient Earth. Could life have started there, too?

For life to begin, there has to be water. Rovers have found direct evidence that water once flowed on Mars. About 3.8 billion years ago, there was water on this planet too. Back then, the Martian canyons may have been filled with chilly water. The southern highlands may have been dotted with icy crater lakes. And geysers near volcanic hot springs may have sprayed water high into the Martian sky.

Carbon, hydrogen, nitrogen, and oxygen —the elements that combine to form organic molecules—were delivered to Mars by icy comets and wayward planetesimals that pelted its surface. These basic elements would have been present in the waters of early Mars, just as they were in the waters of early Earth.

There was also energy available on Mars. Molten lava churned below the rocky surface and erupted through volcanoes and cracks in the ground. Bolts of lightning may have crackled in the Martian sky. Dim but powerful sunlight fell on the Martian surface. One of these might have triggered the building of organic molecules from simpler molecules in the water.

Although Mars was a smaller, more distant world, life could have started there just as it started on Earth. Many of the same watery habitats that nurtured life on Earth may also have existed on Mars. Life may have started at the bottom of a shallow salty sea, where organic molecules collected in the muddy red sediments. Or it may have started in volcanic hot springs, like those in Earth's frigid Arctic, where steaming waters rise to the surface through the frozen ground. Given enough time, organic molecules in these Martian waters could have formed primitive organisms.

But was there enough time? The conditions on the planet quickly began to change.

As Mars' atmosphere began to thin, more of its heat escaped to space. For a while, temperatures sometimes rose above freezing during the day.

Ice-covered lakes may still have held water, insulated by the layer of ice above or warmed by heat from below. Eventually, though, Mars' atmosphere became so thin, and its temperature so cold, that all the water disappeared from its surface. Gone were the icy lakes, the water-filled canyons, and the muddy red rivers. If life did begin on early Mars, could it survive?

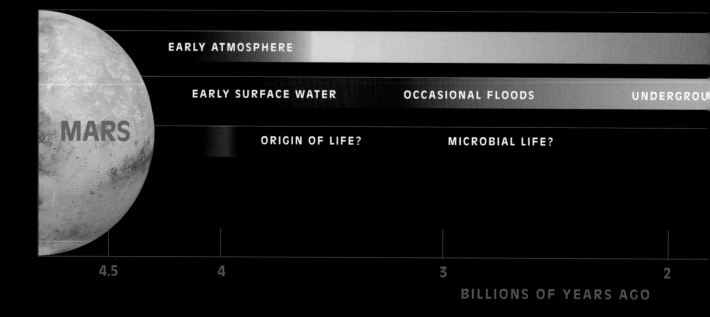

MARS

EARLY ATMOSPHERE

EARLY SURFACE WATER   OCCASIONAL FLOODS   UNDERGROU

ORIGIN OF LIFE?   MICROBIAL LIFE?

4.5   4   3   2

BILLIONS OF YEARS AGO

↑ Painting of Mars as it might have looked about 3.8 to 4 billion years ago.

← Mars time line from 4.5 billion years ago to the present, showing the approximate times of some important periods in Mars' evolution.

TER

1   0.5   now

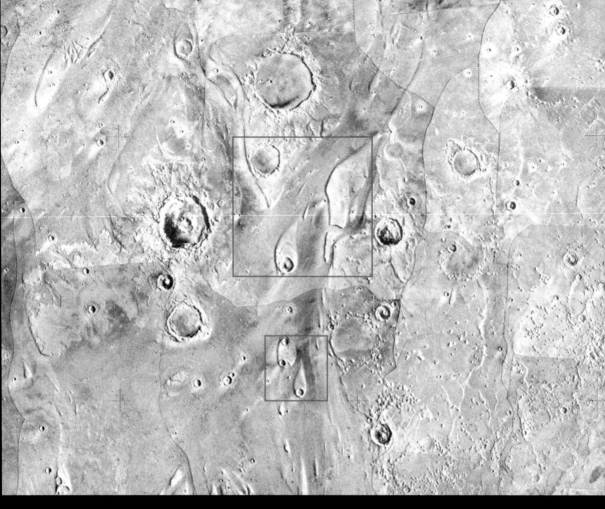

↑ This photo mosaic shows the ancient Martian floodplain where Ares Vallis joins Chryse Planitia. The floodwaters rushed from bottom to top across the middle of the picture. The Pathfinder spacecraft landed in this region. (Viking)

↓ Close ups of landforms in the picture above. Bottom Left: Rushing water carved streamlined islands where it encountered obstacles, like crater rims. Bottom Right: The teardrop-shaped island is nearly 30 miles long and about 2,000 feet high—as tall as a small mountain on Earth. Its crater is over six miles across. (Viking)

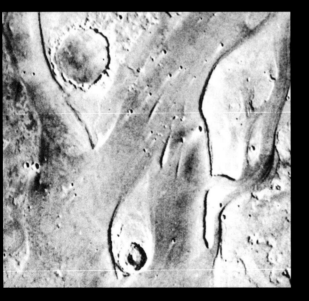

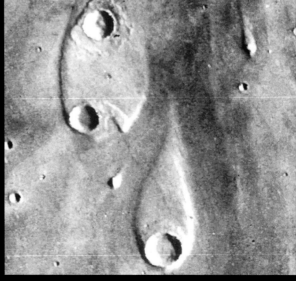

Even after water vanished from the surface, Mars still had lots of water. Some was frozen at the poles, but most of it was hidden underground. The evidence comes from spacecraft photographs.

Some photographs show deep channels and teardrop-shaped islands that were carved by catastrophic floods sweeping across the Martian land. But some of these floods occurred long after the water had vanished from the planet's surface. Scientists believe that impacts by large meteors or heat from rising molten lava suddenly melted ice beneath the ground, releasing giant waves of water.

Over the years, there have been several great Martian floods. This means that a huge amount of water or ice is stored below the surface. It may be miles down, perhaps in huge reservoirs capped by a thick layer of ice. It may be trapped in porous underground rock that soaks up water like a sponge.

↓ *This flat plain near the Martian equator could be a frozen sea. Scientists think that a volcanic eruption or mars-quake sent a huge pool of underground water gushing onto the surface. It quickly froze, then eventually broke into ice sheets and was covered in red dust. Meteor impacts have scooped out craters in the ice.* (Mars Express)

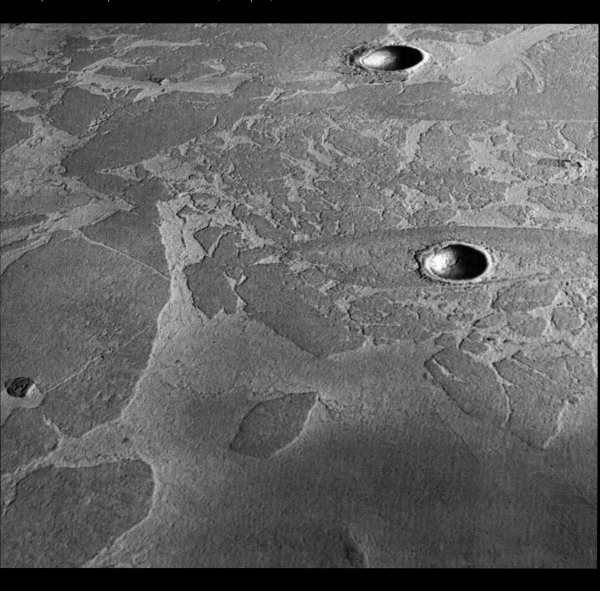

Martian microbes, if they did exist, might have survived underground—in places where there was still water. Primitive organisms are very hardy. Life thrives in some of the most unlikely places on Earth. Colonies of bacteria live in sandstone rock in the frigid deserts of Antarctica; others live in the boiling-hot waters of Old Faithful in Yellowstone National Park; and still others live in almost-solid rock, miles beneath Earth's surface.

Some Martian microbes might have lived in similar, seemingly hostile, environments. Maybe they lived miles deep, in water heated by scalding volcanic vents. Maybe they lived near the surface, in the moist cracks in the rock. If primitive life did begin on Mars, it had a chance to survive as long it was near water.

Throughout the last four billion years, Earth has been a place where life could exist. Its temperature never got so cold that all the liquid water froze, and its atmosphere did not slowly drift away into space or freeze out onto the surface.

Mars is a different story. As time went on, its once-thick atmosphere became thin, and its temperature dropped. Water, the key to life, vanished from its surface. If life did begin on Mars, it could only have survived underground.

Extremophiles

Life on Earth is much more hardy, diverse, and adaptable than scientists used to think. Bacteria live in some of the harshest environments on Earth—places that scientists once thought were lifeless. These bacteria are called extremophiles (which means "lovers of extremes") because they have adapted to live in places that are very hot, or very cold, or very extreme in some other way.

Some bacteria live in the boiling-hot waters near deep-sea volcanic vents. Superheated, mineral-rich water billows out of these undersea chimneys from deep inside the Earth. Heat-loving bacteria grow on their outside walls. These bacteria stop growing if the temperature drops below 200 degrees Fahrenheit —the water's too cold for them!

Other types of bacteria flourish in the coldest places on Earth. Vast parts of the Arctic and Antarctic are frozen solid most of the year. But cold-loving bacteria make their homes in these icy lands.

The strangest extremophiles may be the ones that dwell deep underground, inside almost-solid rock. Trickles of water seep into the rock and keep them alive, but their home is scorching hot and there is no space to grow. Some of these strange bacteria may live for thousands of years.

Before scientists discovered extremophiles, they never imagined that life could exist in such alien environments. On Earth, it appears that wherever there is even a small amount of water, there is life.

↓ *Deep-sea volcanic vents, such as this one more than 10,000 feet below the surface of the Atlantic Ocean, are home to heat-loving bacteria and other "extreme" life forms.*

Life On Earth

For over three billion years, microscopic organisms were the only inhabitants of Earth. And for most of this time, these tiny living things were very primitive. But slowly things began to change.

Some of these microorganisms evolved clever new ways to get nourishment. Eventually, they evolved the ability to use the energy in sunlight and the carbon dioxide in air to make food. This process is called photosynthesis. As part of photosynthesis, oxygen is released into the atmosphere.

By one billion years ago, vast chains of photosynthesizing bacteria formed slimy mats on the banks of streams. Green algae floated near the ocean surface and clung to rocky shores at low tide. Steadily, oxygen bubbled out of the waters and drifted up into the air.

Eventually, about half a billion years ago, simple green plants began to spread across the land. At first, they were very small and could grow only where it was very wet. Later, plants of all sizes and shapes grew farther and farther from the water's edge. But life depends on water. So these plants evolved roots that could drink from underground, instead of having to rely on the morning dew or an afternoon rain.

Once green plants began to cover the land, other forms of life were able to follow. Plants provided food and shelter for other living things. By about 100 million years ago, grasshoppers chewed on young leaves, mushrooms digested fallen logs, frogs hid in the marsh grasses, and birds nested in the branches of bushes. Human beings have inhabited Earth for only a few hundred thousand years—just the last instant of our planet's long history.

Earth today is far different from the barren planet it used to be. Even a handful of soil from a forest is packed with microbes and insects. Even a cup of water from a stream is brimming with bacteria and algae.

↓ Earth

↓ Mars

for centuries. Geologists have collected samples of rock and soil from Antarctica to Tibet. Chemists have analyzed samples of air from around the globe, and biologists have examined ancient fossils from every continent.

It is much harder to study the rocks, soil, and air of a planet that is millions of miles away. No geologist, biologist, or chemist has ever explored Mars. Only robot spacecraft have visited its surface, and all of them are still parked there on the dusty Martian plains. None has brought bags of Martian rocks back to Earth.

Scientists have only the smallest trickle of data from Mars. But that is changing. The launches of *Pathfinder* and *Mars Global Surveyor* began a new robotic assault on the Red Planet. They were followed by *Mars Express*, and the *Mars Exploration Rovers,* among others.

Many more spacecraft will follow these. Every 26 months, when Earth and Mars are favorably aligned in their orbits, spacecraft will rocket toward the Red Planet. The orbiters, landers, and rovers of the future will search for evidence of water and possible primitive life. Some missions will bring precious Martian rocks back to scientists on Earth.

↓ *Launch of* Mars Global Surveyor *aboard a* Delta *rocket.*

Future orbiters will carry high resolution cameras to look at Mars' surface in detail. They'll carry instruments to search for sites where water once pooled or ancient hot springs bubbled. They'll look for interesting places to send future landers.

Microprobes of the future might slam into the planet and tunnel into the ground to look for ice below the surface. Future landers might settle down near the frosty fringes of the polar icecaps, or carry small airplanes designed to catch the wind and soar through the Martian skies. Others will carry larger and larger rovers to explore more of the Martian terrain and look for signs of water and primitive life.

Eventually, a mission will be designed to bring samples of Martian rocks and soil back to Earth. The first mission will ferry about one pound of Martian rocks to eager scientists. These priceless pieces of Mars will be whisked to a quarantine facility to be sure that they do not carry alien microbes that might be harmful to living things on our planet. Then scientists will use sophisticated equipment to study the rocks, soil, and air. They will look for clues to conditions on early Mars and for traces of early microbes.

↓ *Artist's drawing of the Mars* ARES *aircraft.*

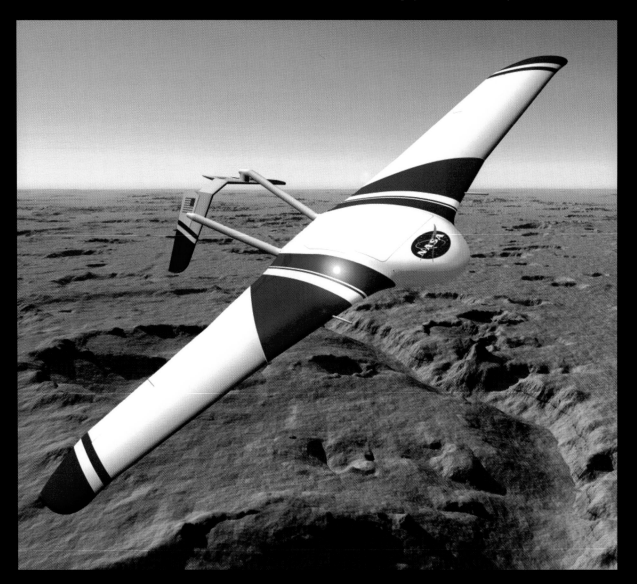

the long journey to Mars. When they step out of their spacecraft onto the red Martian soil, they will become the first human beings to visit another planet. This expedition will extend our presence farther into the solar system. It will be the adventure of a lifetime.

Astronauts will explore Mars the way scientists explore remote areas of Earth. Over many missions, they will plunge shovels into dry riverbeds and chip rocks from steep canyon walls. They will drive buggies over windswept dunes, hike across lava fields, and climb the slopes of extinct

frozen ground for traces of water. If robot scouts have located ancient hot springs, astronauts will scour these sites for rocks and minerals that may contain microscopic evidence of life.

Earth and Mars share a common beginning. But the two planets evolved very differently. Why did Mars follow one path while Earth followed another? As we investigate the mysteries of Mars, we are also learning things that help us understand our planet and how it came to be the oasis that we know today.

Mars Mission Time Line

Mariner 4 (1964 -1965)
Flew past in July 1965 at a distance of 6,000 miles, sending back 22 photographs, which showed craters on Mars' surface for the first time.

Mariner 6 and Mariner 7 (1969)
Both flew within 2,200 miles of Mars, sending back more than 200 photographs.

Mariner 9 (1971)
Orbited Mars, becoming the first spacecraft to orbit a planet other than Earth. Sent back more than 7,000 TV pictures, showing such features as Olympus Mons, Valles Marineris, and Mars' moons, Phobos and Deimos.

Mars 3 (1971)
A probe containing science experiments launched by the Soviet Union. Landed on the surface of Mars but fell silent after 20 seconds of operation.

Viking 1 and Viking 2 (1975 - 1976)
On July 20, 1976, Viking 1 became the first spacecraft to land successfully on Mars. It was followed several weeks later by Viking 2. Both landers sent data to Earth and conducted experiments to search for primitive life. Meanwhile, the Viking 1 and Viking 2 orbiters photographed the Martian surface.

Pathfinder (1996 - 1997)
Landed on Mars on July 4, 1997. Pathfinder sent back thousands of photographs of the surface, and its six-wheeled rover Sojourner became the first vehicle to explore another planet.

Mars Global Surveyor (1996 -)
Orbiting spacecraft designed to survey Mars. Instruments include cameras and the Mars Orbiter Laser Altimeter, which measures the height of features on the surface. Still operating in 2006.

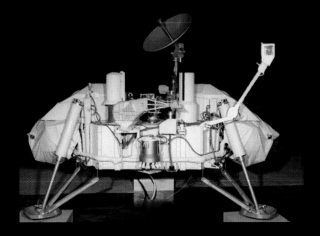

↑ Viking lander.

← Phobos, one of the two moons of Mars. (Mars Global Surveyor)

Distance from sun	128 - 155 million miles	91 - 94 million miles
Diameter	4,219 miles	7,926 miles
Length of year	687 days	365 days
Length of day	24 hours 37 minutes	24 hours
Moons	Two: Phobos and Deimos	One

Mars Odyssey (2001 -)

Launched in 2001, entered orbit around Mars later that year. Its instruments gather information on the Martian climate, geology and mineralogy. Still operating in 2006.

Mars Express (2003 -)

Entered orbit around Mars in December 2003 to search for water-bearing minerals, peer underground, and photograph the entire planet in high resolution. Still operating in 2006.

Mars Exploration Rovers — Spirit and *Opportunity* (2003 -)

The twin rovers *Spirit* and *Opportunity* landed on different parts of Mars in January 2004. Each drove over four miles from its landing site, examining the Martian rocks and terrain. Both found direct evidence that water once flowed on Mars. Still operating in 2006.

Mars Reconnaissance Orbiter (2005 -)

Entered orbit around Mars in March 2006 carrying the highest resolution camera ever sent to the planet. Its instruments provide a detailed view of Mars' geology and help identify future landing sites. Still operating in 2006.

Web Resources

www.nasa.gov
NASA's home page, with current space news and links to other NASA sites.

www.nasa.gov/kids.html
Dozens of links to NASA sites designed for young people.

mars.jpl.nasa.gov
Home page for NASA and the Jet Propulsion Laboratory's Mars missions, with links to other Mars mission web sites.

nsdc.gsfc.nasa.gov/planetary/Mars
National Science Data Center's Mars web site. Includes pages on the history of Mars exploration, a virtual "tour" of Mars, and links to other Mars science sites.

About the Authors

Sally Ride has been interested in science since she was a child. She earned bachelor's degrees in physics and English and a Ph.D. in physics from Stanford University. In 1983 she became the first American woman to fly in space, when she made a six-day flight aboard the space shuttle *Challenger*. She made her second trip into space in 1984. Dr. Ride is now a professor of physics at the University of California, San Diego. She is also founder of Sally Ride Science, a company that creates innovative programs and publications for young people interested in science.

Tam O'Shaughnessy and Sally Ride have been friends since they were teens competing in junior tennis tournaments. Dr. O'Shaughnessy holds a master's degree in biology and a Ph.D. in school psychology from the University of California, Riverside. She is a professor of school psychology at San Diego State University. Tam and Sally have written four other children's science books including *The Third Planet: Exploring The Earth From Space*, winner of the American Institute of Physics Children's Science Writing Award.

Index

Picture Credits

Air Force Chart and Information Center: 8 bottom, ESA: 27 top, 39, Lowell Observatory: